Yellow Umbrella Books are published by Red Brick Learning
7825 Telegraph Road, Bloomington, Minnesota 55438
http://www.redbricklearning.com

Library of Congress Cataloging-in-Publication Data
Catala, Ellen.
   [I love a parade. Spanish & English]
   I love a parade/by Ellen Catala = Me gustan los desfiles/por Ellen Catala.
   p. cm.
   Summary: "Simple text and photos present opportunities to practice counting and
to share in the excitement of watching a parade"—Provided by publisher.
   Includes index.
   ISBN-13: 978-0-7368-6017-8 (hardcover)
   ISBN-10: 0-7368-6017-7 (hardcover)
   1. Counting—Juvenile literature. 2. Parades—Juvenile literature. I. Title: Me gustan
los desfiles. II. Title.
QA113.C38718 2006
513.2'11—dc22   2005025848

Written by Ellen Catala
Developed by Raindrop Publishing

Editorial Director: Mary Lindeen
Editor: Jennifer VanVoorst
Photo Researcher: Wanda Winch
Adapted Translations: Gloria Ramos
Spanish Language Consultants: Jesús Cervantes, Anita Constantino
Conversion Assistants: Jenny Marks, Laura Manthe

Photo Credits
Cover: Long Photography/Tournament of Roses; Title Page: Wernher Krutein/Photovault;
Page 4: Stephanie Maze/Corbis; Page 6: Hillsdale County Chamber of Commerce;
Page 8: Mark Karrass/Corbis; Page 10: Owen Franken/Corbis; Page 12: Nancy White/
Capstone Press; Page 14: Patti McConville/International Stock; Page 16: Stan Ries/
International Stock

1 2 3 4 5 6 11 10 09 08 07 06

I Love a Parade

by Ellen Catala

Me gustan los desfiles

por Ellen Catala

Yellow
Umbrella
Books
for early readers

Come count with me.

Ven a contar conmigo.

How many banners
do you see?

¿Cuántos letreros
puedes contar?

How many horns in a row?

¿Cuántas trompetas
puedes contar?

How many clowns
at the show?

¿Cuántos payasos
puedes contar?

How many flags
way up high?

¿Cuántas banderas
puedes contar?

How many balloons
in the sky?

¿Cuántos globos
puedes contar?

I love a parade!

Me gustan los desfiles.

Index

Índice